Common Medicinal Plants

Uses and Cultivation Practices

Harshita Joshi

Preface

The book "***Common Medicinal Plants: Uses and Cultivation Practices***" explains in detail about some of the commonly grown medicinal plants that can be easily cultivated at home gardens as well as on a commercial scale. The benefit of growing Medicinal Plants at home gardens, apart from their easy cultivation, includes their uses as medicines without any side effects.

Some vegetables and fruits also possess medicinal properties and because of this, all health practitioners in alternative medicines advice to consume fruits and vegetables in large quantities.

The book covers cultivation practices of some of the medicinal plants that are easy to grow. The last part of the book is a quick reference guide in tabular format so that readers can have easy and quick information about medicinal plants at a glance.

Table of Contents

Tables

Introduction to Medicinal Plants

In simple words, medicinal plants may be defined as plants that possess medicinal properties and help in curing various ailments of the body either slowly or rapidly. However in terms of plant science, there are many plants that secrete special chemical compounds from different parts to survive in adverse conditions. The compounds synthesized by plants were identified and now are used for therapeutic purposes in all the herbal industries. It is said that almost all the plants have medicinal properties and hence advised by all the health practitioners to consume all those fruits, vegetables, roots and herbs that are considered edible. These active ingredients synthesized by plants help them to get protected from predators for example animals, microorganisms, rodents, birds and rodents. Following this property, various growers have started the commercial cultivation of medicinal plants on a larger scale to generate profit.

Unique Features of Medicinal Plants

Medicinal plants are recognized and regarded by all the florists and herbal as well as cosmetic industries owing to their potential to cure various disorders. These plants have unique features due to which these are cultivated by both commercial florists and home gardeners. Some of these features are presented here in the next section.

Type of Body Disorders Cured By Medicinal Plants

The naturally occurring compounds secreted by the medicinal plants help in curing all types of disorders of the body that include mental, respiratory, stomach, skin, intestinal etc without any side-effects. In addition, a single plant has the capability to cure multiple disorders. The properties of these plants are also identified and enlisted by World Health Organization. Some of the examples of these medicinal plants may include Indian Neem, Saffron, Anethum, and Turmeric. The detailed descriptions of medicinal properties are given in subsequent sections of the book.

Presence of Active Ingredients

Active ingredients are biologically active chemicals that are synthesized by the plant parts. Some of the examples of active ingredients present in the plants are alkaloids, glycosides, phenols, flavonoids, polysaccharides, saponins, tannins, volatile oils, anthocyanins. The examples of active ingredients and plants are tabulated below:

Table 1: Active Ingredients and Examples of Medicinal Plants

S.N.	Active ingredient	Example of plants
1	Alkaloids	Periwinkle, Acacia, Poppy, Coffee, Tobacco, Datura
2	Glycosides	Foxglove, Digitalis, Lily of the Valley Kalanchoe
4	Flavonoids	Basil, Onion, Garlic, Green leafy vegetables
6	Saponins	Primula, Liquorice
8	Volatile oils	Lemon grass,
9	Anthocyanins	Grapes, Blackberry
10	Quinine	Cinchona
11	Glucosilinates	Mustard, Radish, Most of the crucifers

Source: *WHO monographs on selected medicinal plants VOLUME 3, 2007*

Ornamental and Foliage Trees

There are various plants that can be grown at the garden to serve two purposes. Some of the plants like Lavender and periwinkle that possess medicinal properties also bear attractive flowers hence can be planted in the gardens. In addition, various plants the foliage of which are used as vegetable and consumed either in cooked form or raw, also possess medicinal properties.

Commercial Cultivation

The future prospects of cultivating medicinal plants on a commercial scale are very bright as the crop does not require too much management and produce quality yield. Both the extracted and raw forms of medicinal parts of these plants are purchased by the herbal and cosmetic industries at very higher prices.

Medicinal Plants: How to Use Them as Medicines?

The parts of medicinal plants that secrete appreciable quantities of active ingredients are extracted in herbal drug manufacturing units in order to reap the benefits. The extracted material is then converted into various forms such as powder, tablets, tonics, decoction etc. Beside this, one can also prepare medicines at homes if raised at their gardens by altering the original forms of the plant parts that are generally recognized as having more medicinal effects. The only thing to be taken care of is the proper post harvest management of plants. For example, proper drying of roots in case of plants like Sarpagandha and Ashwagandha or leaves in case of mint. One should have a thorough knowledge of ideal time of harvesting of these plants and after care of the harvested produce to effectively use them as medicines.

In addition there are various plants that are eaten either in raw or cooked form or as a flavoring agent that have medicinal properties. Some of the examples of such plants include ginger, turmeric, all green vegetables, fruits such as papaya, apple etc.

Benefits of Growing Medicinal Plants at Home

Archaeological evidences also prove that plants were the major sources of medicines for human beings of that era. The property of healing the wounds and diseases without any side-effects in addition to easy cultivation practices enables the medicinal plants to be grown in the home gardens. Also, consumption of these plants provides some amount of nutrition to the human body. Some of the major benefits of growing medicinal plants by the gardeners at their home gardens can be summarized in the below points.

No Side Effects

Health practitioners say there are no side effects in the body if these medicinal plants are consumed even in larger amounts as compared to allopathic medicines as these contain chemicals that are synthesized by plants naturally. The medicines manufactured non-naturally can alter the body functions after sometime however plants that already have natural compounds and healing properties can be taken whole life. There are some plants parts of which are consumed as fruits, vegetable and spices; also have medicinal properties. However overdoses can also alter the functioning of the body hence it is strictly advised to consume the medicinal plants after knowing the quantity that can be taken in the form of medicine.

Cheap and Best Source

Medicinal plants if grown in home or kitchen gardens are the cheapest sources to cure one from health disorders. As these are the natural sources of medicines and contain no side effects, medicinal plants can be regarded as the best sources of medicines available in the world for human benefits. An appreciable number of plants have medicinal properties however benefits of many are still unknown.

Easily Cultivable

Proper crop management practices are to be followed if medicinal plants are cultivated on commercial basis otherwise these can be easily grown in the home or kitchen gardens as one grows other vegetables or fruits.

Do you know?

Some plants like Aloe vera, Basil, Thyme, Dill, Mint, Turmeric, Chamomile and Lemon grass are recognized as medicinal plants throughout the world due to their known benefits. Some of these plants are also used as flavoring agents.

Some Common Medicinal Plants

Medicinal plants are consumed in various forms either as a drug in the form of tablet and tonics or as herbs or decoction. Numerous plants are discovered by the scientists that have medicinal properties and there are various plants that are very common to us in the form of either as flowering plant or foliage plant but the medicinal properties are very less known to the people. Even very few people are aware about the medicinal benefits of plants that are consumed in the form of vegetables and flavoring agents as spices. This chapter details medicinal benefits of some of the very common plants grown in the home gardens and in the vicinity.

Aloe vera
Aloe vera is considered as one of the important medicinal plants and is planted by various home gardeners. It acts as a cooling agent against minor burns and skin related problems. It is taken as a cure to anxiety, tension, fatigue, depression, fever, insomnia and headaches however higher doses are always restricted.

Anethum
Anethum or Garden dill is an effective medicine against stomach ache, gastritis and flatulence. It is generally grown in kitchen gardens and used a flavoring agent. The traditional people have used this herb as an aphrodisiac and appetite stimulant. It is also effective against diarrhea, asthma, insomnia, kidney stones and disorders of gall bladder. The fruit has anti-inflammatory properties too.

Bishop's weed
Bishop's weed is a beneficial as a muscle relaxant. Traditionally this medicinal plant was used to treat gastrointestinal cramps, asthma, diabetes, stones in the kidneys, and as an antispasmodic. Spanish carrot is another name for Bishop's weed.

Clove

Clove is a used as a flavoring agent that has medicinal properties especially against toothache, or in other words it has anesthetic properties. It is highly recommended not to use the flower bud in large amounts as it has various side effects like vomiting, diarrhea, throat problems, intestinal bleeding, and increase in rate of heart beat. High consumption can also cause kidney failure.

Fennel

The dried fruits of fennel are used in herbal industries as it is a good digestive agent. Other major uses of fennel include treatment against general stomach disorders like flatulence, constipation, and gastritis; treatment against fever, diarrhea, headache, pain and poor appetite. It also has aphrodisiac properties.

Fenugreek

Another common name for fenugreek is Greek hay. It is an annual herb that is grown in kitchen garden for its leaves and dried seeds. Another species of fenugreek is used as a flavoring agent as it contains aroma. It is clinically proven that the seeds of the plant can be used against diabetes. Other medicinal uses of fenugreek include its uses as an appetizer, pain reliever, weaknesses, and diuretic. It is also a good medicine to cure bronchitis, diarrhea, gout, indigestion, fever, cough, liver disorders, and common cold.

Ferns

The juice of soft leaves of Lady Fern has medicinal properties to treats minor cuts, burns, and stings. The juice of this plant is very effective for the treatment of burns by stinging nettles.

Ginger

Ginger is a spice crop that also has therapeutic properties and is used by many herbalists for manufacturing herbal drugs. The major medicinal benefits of ginger are treatment against cold, stomach disorders, motion sickness, anti-inflammatory, bronchitis, nausea, and weak immune system. Studies have also shown that consumption of ginger helps in fighting deadly diseases like cancer.

Indian Neem Tree

It is also known as China tree however is indigenous to India. The height of this tree ranges from 5 to 25 m and almost all the parts are used as medicines. Neem is also used as botanical and the details are described in this book in other section. Neem is used as an effective cure against ringworm, lice, malaria and external ulcers. Neem is also considered as a remedy to jaundice, skin disorders, heart disorders, and boils. In traditional times, neem is also used by the people to treat asthma, headache, and stones in kidneys, yellow fever, and smallpox. As a whole, neem can be used medicinally against all the body disorders.

Lavender

Lavender is also grown by many of the home gardeners for its beautiful flowers. Commercially lavender is cultivated for its oil. The height of this shrub ranges between 1 to 2 m. The oil of lavender has medicinal benefits and it is clinically proven that the inhalation of lavender oil is used as a treatment against anxiety, and restlessness. It is also an effective medicine to treat sleep disorders, gastrointestinal disorders, headaches, diarrhea, wounds and throat problems.

Mint

There are various species of mint that are grown commercially for their aromatic properties as it fetches good amount of returns. Mint oil is used in herbal industries for the preparation of various medicines/syrups. Mint contains menthol and is used as a treating agent against headaches, nausea, stomach ache, cold and cough as it has anti-viral properties. It is also used as mouth freshener against bad breath. Other benefits of mint include relieve against fatigue and nervousness.

Papaya

The fruit, seeds, stems, and leaves of papaya have medicinal properties. The proteolytic enzyme present is papaya help in digestion of food. Papaya is very effective against cataract, cancer and heart disorders.

Passion flower

Passion flower is a natural relaxant and is used to treat insomnia and anxiety. It is a mild stimulant that is used to treat against various gastrointestinal disorders. Passion flower also has anti-inflammatory properties.

Common Periwinkle or *Vinca rosea*

One of the very beautiful flowers, periwinkle is grown for its ornamental as well as medicinal properties. It is one of the very important medicinal plants used as a cure against blood cancer. As a medicine periwinkle is used to increase blood circulation and memory. Other benefits also include treatment against diarrhea, throat problems, hypertension, toothache and inflammation. In other words, periwinkle has anti-bacterial, anti- cancer, diuretic and sedative properties.

American Saffron or Indian Safflower

Indian Safflower is an annual herb that is used as a medicine against wounds, sores, pain and swellings. The medicinal plant is also effective against fever, diarrhea, respiratory infections, scabies and sometimes ringworms. Other benefits of this herb include its sedative and laxative properties.

Zizyphus or Chinese Date

Also known as Jujube, Chinese date is a spiny shrub. It is a deciduous shrub and the height of the plant reaches not less that 9 to 10 m. The fruits of this shrub are used as medicines to cure insomnia, asthma, bronchitis, eye infections, ulcers, scabies and wounds. The consumption of Chinese date also helps in gaining weight and muscular strength.

Easily Cultivated Medicinal Plants for Kitchen Gardens

Besides these common plants that are grown by almost all of gardeners in their home and kitchen gardens, there are some plants that are only grown because of their medicinal properties. This chapter presents some of the important plants that can be grown easily in the home gardens for therapeutic purposes. There are various flowering and foliage trees that also have medicinal properties too. One should have fair idea about the medicinal plants and their uses so that medicinal parts of these can be taken in place of allopathic medicines. However utmost care should be taken before consuming these plants as medicines as there are medicinal plants grown at the home gardens but the recommended quantity of medicinal part that is to be consumed is not known to most of the people. Hence overdoses can also lead to side-effects because the chemical compounds synthesized by the plant can alter the body functioning if consumed above the recommended quantity. This chapter along with the cultivation practices of some of the important medicinal plant will also cover the effects of overdoses from the advised quantity.

Ashwagandha

Introduction

It is also known as Indian ginseng. The botanical name of Ashwagandha is *Withania somnifera Linn.* And it belongs to the *Solanaceous* family. The seeds, leaves, and roots of the plants are used in many herbal industries for preparation of various medicines. As a medicine, ashwagandha is used for the treatment of arthritis, insomnia, asthma, bronchitis, and liver disorders. Other medicinal benefits of ashwagandha include treatment against ulcers, eye infections, joint inflammation, and female disorders.

Cultivation Practices

The ideal soil for growing ashwagandha is sandy loam with the pH range of 7-8. The soil should have proper drainage and good amount of organic matter. Ashwagandha is planted in late rainy season generally in the months of August and September. Due to its various medicinal benefits, commercial cultivation is in practice however it can be grown in home gardens also. The plant is propagated by seeds but first seedlings are raised in nurseries. Other methods of propagation also include direct sowing in the field by broadcasting. The nursery method is however preferred over broadcasting. Nursery is raised in the months of June and July and soil should contain a good amount of organic matter for quick germination of seeds plus sand for proper aeration. It takes around 5-8 days for a seed to germinate. Before sowing the seeds, proper crop management practices should be followed such as dipping the seeds in an appropriate fungicidal solution to prevent fungal infection of seeds. Light watering is done to keep the plantings healthy. Thinning is also carried out on the main field when seeds are broadcasted directly. The seedlings are transplanted after 35 to 40 days of their germination.

Before planting on the main field, the land is prepared by deep plowing and leveling. While planting at the home gardens, deep hoeing with addition of organic matter is advised. The nursery seedlings are sown in the main field and proper plant distance ranging from 30 to 45 cm is maintained to carry out intercultural operations. The plant responses well with the addition of organic manure and chemical fertilizers are not applied. Heavy irrigation is not recommended as it can damage the plants and as the planting is done during rainy season, there is no need of irrigation or light irrigation, if required.

The plants are ready to harvest after 150 to 180 days of planting on the main field generally in the months of February- March. One can identify the ideal time of harvesting by looking at the dried leaves and mature berries. Whole plant is uprooted as roots have therapeutic properties too. The roots are cut and fried under the sun. There is a proper method of cutting and drying of the roots too. In case of fruits, the process starts from hand plucking, drying and crushing to separate the seeds.

Aloe vera

Introduction

Botanically Aloe vera is known as *Aloe barbadensis* Linn and it belongs to the family *Liliaceae*. The medicinal properties of Aloe vera have been already discussed in the earlier sections.

Cultivation Practices

The cultivation practices of Aloe vera are simple. It grows best in the dry and hot climate and can survive harsh conditions. The type of soil ranges from heavy hill and black soil to light loamy and is propagated by suckers. The plants at the home gardens can also be grown as a potted plant and if planted on the field, the land is thoroughly prepared by lightly ploughing the soil that is followed by leveling. At the time of soil preparation, addition of well decomposed farm yard manure help in boosting up the growth of plants. Rainy season is the ideal time of planting Aloe vera and the spacing between the plants and rows should be in the range of 45 to 60 cm. Irrigate the crop once in a fortnight during summer season planting while no watering is required if the crop is cultivated in rainy season.

Addition of recommended quantity of fertilizers also enhances the growth of Aloe vera. In addition, other crop management practices such as pest control can also be taken care of. This medicinal plant is prone to diseases such as Brown spot hence pest management practices are also required.

The leaves become ready to harvest after one year of planting. The leaves are cut with as sharp knife and avoid uprooting the complete plant. The viscous gel of the plant is used by the herbal drug manufacturers and cosmetic industries.

Basil

Introduction

Botanically, basil is known as *Ocimum basilicum* and it belongs to the family *Lamiaceae*. Basil is also known as Sweet basil and Tulsi in some of the regions. More than 50 species of Basil are cultivated all over the world however the plant is widely in cultivation in some of the countries like India, Egypt, Indonesia, France and United States. The plant is an erect bushy herb of about 60 to 90 cm and with maturity, the soft stem of the plant also becomes woody.

As a medicine stem, leaves, seeds and flower tops are used. In addition to their use a medicine these parts are also used as a flavoring agent. The medicinal properties of treatment against common cold, sore throat, fever, cough, stress, heart, skin and teeth disorders, and relieve from stomach pain.

Cultivation Practices

Basil grows well in sub-tropical climate and the ideal range of temperature for proper growth and flowering is 10 to 27 degree Celsius. A well drained soil rich in organic matter or humus with good water holding capacity and pH range of 4.5 to 8 is considered ideal for growth of basil. In addition, the soil should also have proper drainage. Basil doesn't grow in drought prone areas.

The plant is propagated by seeds and for appropriate and uniform growth and the seeds are first raised in seedling trays and then transplanted to the main field. The plants can also be raised in pots when space is a constraint. Sunlight also affects the growth and development of Basil. Other than seeds, the plants are also propagated by stem cuttings. The planting distance should not be less than 30 cm and the depth up to 5 mm. the ideal season of planting Basil is rainy season from August to October. The plant does not require fertilizer when soil is highly fertile as the source of nutrients is the decomposed and ready to intake organic matter mixed with the soil during field preparation. Proper irrigation and pest management are other pre-requisites for good growth of the herb.

The plants are harvested as per the requirements. Leaves can be harvested anytime. For extraction of essential oil, harvesting is done once the plants complete their flowering stage and the entire foliage is advised to be harvested before flowering.

Chamomile

Introduction

The medicinal properties of Chamomile are well known since ancient times. It is beneficial towards various disorders of the body and also has nutritional properties. The botanical name of medicinal Chamomile is *Matricaria chamomilla* L. and it belongs to the family *Asteraceae*. It is an annual plant that reaches to the height of approximately 80 cm. The plant can be easily grown in the home gardens as well as commercially on a larger scale as there is an increasing demand of Chamomile and the market is very huge owing to its uses in herbal and cosmetic industries. Chamomile was first originated in Eastern and Southern Europe.

Some of the medicinal benefits of Chamomile include treatment against stomach disorders such as flatulence, stomach ache, diarrhea, indigestion. Chamomile is an antiseptic that is beneficial against wounds, skin problems and infections. It also has anti-inflammatory properties. In addition, in cosmetic industries, the extracts of Chamomile is used to prepare creams, lotions, perfumes, and in aromatherapy.

Cultivation Practices

Chamomile generally requires heavy and moist soil that should be rich in organic matter however due to its wide ranging adaptability, the plant thrives well is any type of soil and other climatic conditions such as temperature range from 2 to 20 degree Celsius and altitude of 300 to 1500 m. Chamomile is propagated by seeds and seeds are first raised in nurseries for better germination and health. Nursery planting in done in September- October and the plants are ready to be transplanted in the months of November-December. A nursery bed should always be prepared by mixing well decomposed farm yard manure. It is also advised to mix organic manure to the soils that are poor in fertility. The spacing between transplanted seedlings should not be less than 20 cm for better yields of flowers and oil. Chamomile requires irrigation at proper intervals to maintain moisture level up to the optimum. It is also proved that watering the plants during their flowering also help in increasing the yield of flowers.

Other crop management practices include regular weeding and hoeing, and application of recommended quantity of fertilizers that is approximately 50 kg/hectare. Application of fertilizers helps in increase in vegetative growth and overcoming nutrients deficiency.

The flowers are ready to be harvested after three to four months of transplanting. The peak flowering months are March and April and the ideal stage of harvesting is when the flowers are near to full bloom stage. Flowering is a labor intensive operation and it is advised to have a clear cut idea of the harvesting stage in order to get quality and best yield. The plant produce profuse flowering but temperature affects number of flowers.

Turmeric

Introduction

Turmeric is one of the very important medicinal plants used by herbal as well as cosmetic industries. In India, the powdered form of turmeric is used as a spice to add color and flavor to the dishes. The botanical name of turmeric is *Curcuma longa* Linn. belongs to the family *Zingiberaceae* and originated in South East Asian region. It is perennial herb generally 60 to 90 cm in height and is grown for its rhizomes.

The use of turmeric as a medicine is very wide. The naturally synthesized compounds in the rhizomes have properties to block cancer, strengthen the immune system of the body, cure liver diseases, prevent diabetes, and cure diabetes. In addition, it also has anti-inflammatory, anti-oxidant, and anti-septic properties. It is also widely used by cosmetic industries for manufacturing of creams and lotions. Other traditional benefits of taking turmeric as a medicine include treatments against constipation, problems in kidneys, eczema and nasal jam.

Cultivation Practices

Turmeric thrives well in a climate that is moist but hot. The soil should be well drained, fertile with richness of organic matter to boost up the growth. It is also grown as a rain-fed crop in some parts of the world especially where rainfall is high. Turmeric is grown a mixed crop in general with other main crops such as paddy and sugarcane.

The plant is generally propagated by rhizomes as seed propagation is not considered economical. The field is prepared by one deep plowing and leveling. Mixing well decomposed organic matter to the soil increases its fertility. Planting material is prepared by cutting the mother rhizome into thin pieces. The ideal season of planting is summer season and the spacing between the plants should be in the range of 30-35 cm and for ridge planting, 45 X 30 cm. Water the plants once in 5 to 6 days when the soil is heavy as it has good water holding capacity however the frequency of number of irrigations increases as per the soil.

In addition to decomposed organic matter, the rhizome also response well by the application of Nitrogenous, phosphate and potash fertilizers. Mulching also helps in conserving the soil moisture and reducing weed growth. Dry forest leaves and cover crops are good sources of mulches.

The rhizomes are harvested when entire foliage dries up generally after 180- 240 days of planting. Rhizomes are dug with the help of axe or hoe, cleaned and separated.

Lemon Balm

Introduction

In botany, Lemon balm is known as *Melissa officinalis* and it belongs to *Lamiaceae* family. Lemon balm is native to Mediterranean and West Asian region. It is a fragranced perennial herb that is grown for its leaves and oil that have multiple medicinal properties that include its antiviral, anti-cancer and tranquilizing properties. Other important medicinal benefits of growing lemon balm at the home gardens are its effect to treat stomach disorders, insomnia, anxiety, fever, indigestion and headaches.

Cultivation Practices

Lemon balm is a fast growing herb and generally considered as a weed by most of the gardeners. It is propagated by seeds and sometimes cuttings. The soil should be well drained, sandy loam and rich in organic matter with the pH range of 4.5 to 7. The location should be sunny and the ideal season of planting is spring. It can also be cultivated as a container plant. The distance between plants should be in the range of 30 cm and the row to row distance should not be less than 150 cm. Addition of fertilizers enhances the vegetative and flowering growth of the plant. Water the plants at regular intervals too. Thinning the plants is also one of important crop management practices in Lemon balm.

The leaves are harvested as per the requirements and the stems are harvested when the plants started turning yellow. It is advised to harvest the plant in the morning hours. The top one third portion of the plant with flowers contains maximum oil and produce higher yields when harvested during rainy season. The leaves are dried by hanging and stored for future uses however the fresh leaves and stems contain maximum flavor.

Lemon Grass

Introduction

The oil of lemon grass is used in aromatherapy and the medicines prepared from its leaves are used to treat various disorders of the body some of which include stomach disorders, common cold, nausea, hypertension, fever, head ache and muscle pain. Lemon grass is also used as a flavoring agent in food and beverage industries. Other uses of lemon grass also include manufacturing of soaps, and natural citral.

Lemon grass is a tall and perennial grass that was first originated in South East Asian and Australian regions. The botanical name of this grass in *Cymbopogan flexuosus* and it belongs to the family *Poaceae*.

Cultivation Practices

Lemon grass is a plant of tropical and sub-tropical region and climatic conditions affect the amount and quality of oil in the grass especially temperature and rainfall. The grass also requires sufficient sunlight for its growth and development. Soil with good drainage, sandy loam texture and water holding capacity including good amount of organic matter is considered ideal for lemon grass. Propagation is generally done by seeds during rainy season and it takes 4-5 days for seeds to germinate. It is recommended to mix well decomposed organic matter and neem cake with the nursery soil and continuously water the seedlings; for their better growth. The seedlings are raised in nurseries before planting on the main field. Usually the seedlings are planted on the ridges in the area that receive high rainfall. The planting distance should not be less than 40 cm. Irrigate the plant at regular intervals. Weeding, addition of fertilizers, and prevention of attack of insects and diseases are some of the important crop management practices to be followed on regular basis.

Flowering occurs after 120 -150 days of transplanting and the plant is harvested by cutting the plants with sharp blade or sickle leaving 10 cm portion above the ground. The harvested plants are left to wilt before distillation. The oil content varies as per the age of plant. As Lemon grass is a perennial grass, the oil yield is highest in the third year.

Liquorice

Introduction

Botanically Liquorice is known as *Glycyrrhiza glabra* Linn. that belongs to the family *Papilionaceae*. It is a sub-Himalayan perennial herb much is use by herbal industries for manufacturing of drugs that are used to treat sore throat, cough, ulcers, arthritis, fatigue, depression, hypoglycemia, and hormonal imbalances. Roots contain the active ingredient and used by drug industries to manufacture medicines. The compound containing active ingredient that has medicinal properties in Liquorice is Glycyrrhizic acid. The shrub reaches to the height up to 1m.

Cultivation Practices

Liquorice can thrive well on wide ranges of soil that have pH ranges of 4.5 to 8. Forest soil that is rich is organic matter, optimum temperature of 25 degree Celsius in summer, average rainfall up to 75 cm, and proper sunshine are other pre-requisites for a healthy root system of the plant.

Liquorice is propagated by healthy root cuttings during spring or rainy season. Each root cutting should have 2-3 eye buds. The garden is prepared by deep hoeing followed by leveling and removing of the clods. Addition of well decomposed farm yard manure is beneficial. The soil should always have proper water drainage capacity as water logged soil harms the crop. Cuttings take around 15 days to sprout and then planted on the main bed. The planting distance for the main crop should be in the range of 45 to 90 cm. Regular weeding, irrigation during spring planting, addition of green manure and neem cake are other prerequisites for better crop.

As the plant parts used for making medicines are roots, well developed and healthy roots can be obtained after 2 to 3 years of planting. The roots are dug, cleaned, and dried under the sun.

Lavender

Introduction

Lavender is an evergreen shrub that is grown by many of the home gardeners for its attractive flowers. It is easily cultivated plant that does not require much crop management practices. The medicinal properties of this shrub are already discussed in the earlier sections and this section will detail about the growing practices of lavender. The botanical name of lavender is *Lavandula angustifoia* and it belongs to the family *Lamiaceae*.

Cultivation Practices

Lavender is a summer season crop hence the ideal time of planting is the onset of summer generally during the months of April and May. The soil should be well drained, rich in organic matter and preferable sandy loam as the plants doesn't attain its full growth if planted on heavy soils. Lavender can also be grown as a pot plant. The planting distance should not be less than 90 cm to get perfect flowers. Propagation can be done either by seeds or softwood cuttings. The cutting from the young plants are selected and planted on the field or the seeds can be directly sown on the field. Care is taken to water the field at regular intervals however lavender is a drought tolerant crop and can tolerate dry summers too.

It is important to regularly prune the plants by removing the dead and diseased branches and leaves to prevent the spread. Other crop management practices also include regular weeding of the area that surrounds the plants. Some of the gardeners also plant lavenders as a hedge. For that, the planting distance should be in the range of 25 to 30cm.

The flowers stalks are harvested when flowers start showing off the color but not fully open. These stalks are then placed in a cool and aerated place for drying.

Thyme

Introduction

Thyme is cultivated as a spice crop in various parts of the world however the herb also possesses many medicinal properties. In Botanical terms, thyme is known as *Thymus vulgaris* L. and it belongs to the family *Lamiaceae*. The herb was first originated in the Mediterranean region. The active chemical compounds secreted by the herb contain healing and health promoting properties. The herb is generally used to treat stomach disorders, cough and cold, diarrhea, mouth problems and skin disorders. Thyme is also considered as an appetizer. In addition, the nutrient contents of thyme especially Vitamins are also appreciable.

Cultivation Practices

The cultivation practices and crop management practice of thyme are very easy. The herb requires dry and not too fertile soil that is well drained and rich in organic matter. Propagation is done by root splitting. The cuttings are planted at the plant spacing of 30 cm and the spacing between the rows should not be less than 60 cm. Other methods of propagation also include seed and cuttings. The planting is done during the months of March and April.

Once sown, thyme doesn't require much cultivation practices. The plants are watered if necessary when the weather is dry. Mulching is also practiced in some cases. Harvesting is done before flowering and it occurs round the year however the best flavor is obtained in the months of June and July. The shoots are cut and spread for drying and to retain the color it is recommended to dry them in shade or dark. Once dry, the leaves are then separated and stored for future uses.

Medicinal Plants as Botanicals

There are various plants the extract of which can be used to prevent the attack of pests in the field. These plants are known as botanicals. In other terms, the pesticides that are derived from plants are known as botanicals. The positive point to use botanicals for pest control is that these are safer to the environment even if used in large amount in contrast to synthetic chemicals as these are naturally originated and extracted from plants. There are plant species that are beneficial to humans as medicines and also used as botanicals to cure pest attack. One of the best examples for a botanical pesticide is Indian neem that can be used in both ways. As a medicine, Indian neem is used to cure almost all of the disorders of body and as a botanical pesticide all parts of the tree are effective against various pests such as nematodes and other insects, and diseases such as powdery mildew, black spot, and anthracnose. Other examples of botanical pesticides include fennel, neem, garlic, wild marigold, pyrethrum and tobacco.

Disease Pest Management in Medicinal Plants

Introduction

Although medicinal plants have tendency to overcome the attack of certain pests that harm the crop badly with the secretion of chemicals, there are certain insects and diseases of which the management is necessary as if not controlled, it affects the whole crop or garden seriously. This section will describe some of the common pests that attack almost all the medicinal plants and various management measures to overcome from the same. In addition, this section will also deal detail some of the common crop management practices that are required while cultivating any type of plant in the garden or commercially on a larger scale.

Major Insects and Pests of Medicinal Plants

Termites

Termites attack the trunk and root portion of the plant and dries the plant completely. The medicinal plants that are affected by termites severely are Caesalpinia species or Brazil wood, Clerodendrum species, and Commiphora or Guggal. The best control measure to manage the termites are drenching of the land with an appropriate termicide such as Chlorpyriphos. Other control methods also include breaking up of termite mounds and planting of species that have termicidal properties such as Acacia species, Leucaena leucocephala etc.

Rodents

The parts of some of the medicinal plants such as Kalimusli (Curculigo orchioides) and Desmodium are eaten by rats. Rats eat either underground parts like rhizomes or outer portions and thus damage the plant. The attack of rodents is controlled by following standard measures of control such as traps, application of rodenticides etc.

Aphids

Aphids are serious and very common pests of all the horticultural crops including medicinal plants. These insects feed on the cell sap of plants especially from the flower, leaf and stem region thereby de-coloring the portions. The severely affected plant dies away due to deficiency.

Aphids are controlled by removing all the affected plants, force irrigation to the affected portion of plant to wash them away and by regular check on the plants. In addition there are plants that can be planted as border crops which are disliked by aphids such as garlic, chilli, mint and onion.

Locusts

Locusts are one of the notorious insect that eat away the leaf and stems thereby damaging the crop as these insects appear in swarm and affect the crop seriously. Locusts lay eggs on the leaves of medicinal plants and nymphs feed the plants by chewing the parts. Mostly plants of African and Asian countries are affected by locusts.

Some of the major and common control measures include application of bio-pesticides such as neem cakes, parasitic wasps, and removal of affected plant parts. Application of any suitable insecticides is also advised but seldom used in medicinal plants.

Major Diseases of Medicinal Plants

Damping off

Damping off is one of the major diseases of medicinal plants that are raised in nurseries before transplanting to the main field. The disease is characterized by the failure of seedlings to emerge and presence of white fungal growth. High humidity and less air increase the growth of soil borne fungi especially Fusarium and Phytophthora. The disease is controlled by properly raising the nurseries, removal of dead and diseased seedlings, thinning, and spraying of fungicides at recommended doses.

Black rot disease

Black rot disease in medicinal plants is mostly prevalent in African and Asian countries. The disease is identified either before or after harvesting of the plants however the young seedlings also produce symptoms that are identical to other diseases' symptoms. Black rot is caused by bacteria but the plant dies due to secondary infections. The disease attacks the plant in any of its growth stages and is characterized by yellow margins of cotyledons followed by dropping off of the leaves. The symptoms first appear in a small "V" shape leading to black colored veins when the severity increases. The internal organs of the plants also turn black at later stages. It is a fungal disease.

The control methods include the use of disease resistant seeds of the medicinal plants, deep plowing of the field, proper thinning of seedlings in the nurseries, keeping the weed free field and removal of affected plants and parts. Hot water seed treatment is also practices by some of the growers.

Leaf spot

The plants attacked by leaf spots are identified by the presence of round to oval spots on the upper surfaces of the leaves that are followed by the white papery texture and brown margins. Various spots later collapse and form large lesions on the leaves leading to extreme dryness and discoloration of leaves. Leaf spots on medicinal plants are caused by bacteria as well as fungi and the severe attack leads to considerable loss of the yield.

Control measure for leaf spots include removal of dried and affection portion of leaves and sometimes whole plants to prevent the spread. In addition, application of Bordeaux mixture at the rate of 10 ml per liter of water can also prevent the emergence and spread of leaf spot.

Leaf blotch

Leaf blotch is characterized by irregular blotches on the surfaces of mature leaves. The affected leaf turns yellow in due course of time and die. In turmeric, the disease first appears on lower leaves generally during winter season. Numerous small yellow spots cover the entire leaf on both the surfaces followed by reddish brown color and finally abscission. Leaf blotch is caused by fungus hence the control measures are application of any recommended fungicide like Mancozeb at the rate of 3 gm per liter of water. Other crop management practices also include removal of disease affected plant parts and crop rotation once in two years.

Rhizome rot

Rhizome rot is a serious disease of medicinal plants that are propagated by rhizomes such as ginger, and turmeric. Soft rot is another name for Rhizome rot and is a very serious disease that leads to devastation of entire cropped area. The attack of Rhizome rot also affects the total cost of cultivation of medicinal plants. The symptoms start with yellowing of tips of lower leaves that increases downward slowly. The yellowing starts from margin and thereafter covers the entire leaf leading to withering and drying of leaves. The rhizome also shows discoloration with the spread of disease.

Rhizome rot occurs mostly in the areas that receive heavy rainfall and it is difficult to manage once it infects the plant. Crop management practices such as selection of disease free rhizomes, proper drainage facility, removal of affected plant parts, and application of Bordeaux mixture at the rate of 10 ml per liter of solution.

In addition to prevent the infestation of insects and diseases, one should also follow proper management practices to control the emergence and spread of weeds. As weeds compete with the main plants for air, light, space and nutrition and affect the yield and quality of crop to a larger extent, regular weeding, hoeing and mulching of the area where the crop is not planted, is essential.

Common Crop Management Practices

It is important to follow proper crop management practices to get the good quality and yield of crop. Although medicinal plants do not require much attention as the plants secrete chemical compounds that help them to survive in adverse environmental conditions, there are some of the common practices that can be regularly followed to maintain the higher yield of medicinal plants. These common practices are discussed below:

➤ Regular weeding and hoeing of the interspaces between the plants that helps to check the weed growth. Weeds compete with the plants and take away nutrition and other elements that are essential for the plants' growth hence regular weeding helps in minimizing the emergence of weed plants.

➤ Mulching helps in conserving the soil moisture and prevents the growth of weed plants. Suitable mulch like straw, polythene sheets, compost, leaves and cover crops are spread on the interspaces that help in proper growth and development of the main crops.

➤ Mixing of well decomposed farm yard manure is also one of the important crop management practices. Un-decomposed organic matter invites disease causing organisms and lead to the infestation of diseases. Severe infestation of diseases directly affects the yield of main crop resulting in huge economical losses.

➤ Soil treatment before sowing the main crop on the field is also an important crop management practice. Examples include deep plowing and turning of the soil to kill soil borne pathogens and insects by exposure to the sunlight, soil fumigation with formaldehyde at recommended doses, application of suitable pesticides (termicides, rodenticides, and fungicides) to kill the disease causing microorganisms, and soil reclamation.

➤ Removal of affected plant parts in order to prevent the further spread of diseases in the affected area. It is also advised to throw away the affected portion and sometimes the entire plant and burn them.

➤ Application of beneficial biological organisms in the main crop area like birds, certain bees and wasps, and microorganisms also help in increasing yield and quality by checking on pest infestation.

➤ Integrated Pest Management is an approach to sustainable agriculture that helps in controlling the pest economically. Various components of IPM include cultural practices, regular monitoring, mechanical control methods, biological control methods, and at last, use of chemical methods if the attack of pests is uncontrolled.

Commercial Cultivation of Medicinal Plants for Profit Generation

Medicinal plants are required by the herbal industries for manufacture of medicines in various forms such as tablets, tonics and powder. Due to synthesis of chemical compounds in their natural forms, the medicines prepared by medicinal plants do not have any side affects hence relied by majority of the population. In addition, the crop management practices to cultivate the crop on a larger scale is not very difficult as some of these medicinal plants also act as barriers against pests that affect other ornamental and vegetable crops to a larger extent and sometimes completely devastate the land. The medicinal plants in the market are sold either in raw or extracted forms and the prices may vary according to the potential of the plants and semi-finished products. It is always beneficial to cultivate medicinal plants on a commercial basis if land is available. These medicinal plants can be grown on a wide range of soils and climatic conditions also add value to their commercial cultivation.

Some of the medicinal plants that have more potential and can be commercially cultivated on a larger scale in order to generate profit are Ashwagandha, Lemon grass, Mint, Aloe vera, Pyrethrum and Turmeric.

Annexure

Table 2 : List of Medicinal Plants with Their Common Names

S.N.	Name of the Plant	Botanical Name	Other common names
1	Aloe vera	*Aloe barbadensis*	True aloe, Indian Aloe
2	Bishop's weed	*Ammi majus L*	Bullwort, Mayweed, Devil's carrot
3	Dill	*Anethum graveolens L.*	Anethum, Koper, Sowa
4	Apricot	*Prunus armeniaca L.*	Chuli, Wild Apricot, Aprikose
5	Arnica	*Arnica montana L.*	Leopard's bane, Fallkraut, Mountain tobacco
6	Indian Neem tree	*Azadirachta indica A.*	China berry, Indian lilac, Limbado, Margosa
7	Indian safflower	*Carthamus tinctorius L.*	American saffron, Biri, Hong- hua
8	Saffron	*Crocus sativus L.*	Safran, Kesar, Spanish saffron
9	Fennel	*Foeniculum vulgare Mill.*	Fenchel, Marui, Sanuf
10	Gummi Gugguli	*Commiphora mukul*	Guggal, Hill mango, Ranghan
11	English Lavender	*Lavandula angustifolia Mill.*	Common Lavender, Hanan, Al birri
12	Passsion Flower	*Passiflora incarnate L.*	May Apple, Water Lemon, May Flower
13	Indian Plantago	*Plantago ovata Forsk.*	Esopgol, Psyllium, Ghoda
14	Irish Daisy	*Taraxacum officinale*	Blowball, Dandelion, Gol Ghased
15	Fenugreek	*Trigonella foenumgraecum L.*	Bockshornklee, Gandhaphala, Menthi

Source: WHO Monographs on selected medicinal plants VOLUME 3, 2007

Table 3 : Medicinal Benefits and Main Part of Plants Used As Medicine

S.N	Name of the Plant	Botanical name	Medicinal Benefits	Part Used as medicine
1	Ashwagandha	*Withania somnifera*	Against Rheumatism, insomnia, asthma, bronchitis, and liver disorders	Roots, leaves, seeds and fruits
2	Asparagus	*Asparagus racemosus*	Antiseptic, Dysentry	Spears
3	Atis	*Aconitum heterophyllum*	Anti-diabetic, Antipyretic and against irregular menstruation	Seeds and tubers
4	Bach	*Aconitum ferox*	Against Cholera, Rheumatism, and Leprosy, has Antipyretic properties too	Roots
5	Datura	*Datura alba*	Against Cough, inflammation, and Pain	Roots
6	Dolu	*Rheum emodi*	Astringent	Rhizome, Root
7	Garlic	*Allium sativum*	Against Stomach disorders, facial paralysis and Heart disorders	Roots
8	Gurbach	*Acorus calamus*	Diuretic, Brain tonic	Rhizomes

33

9	Isabgol	*Plantago ovata*	Against Dysentry, Diarrhea, Cough, Cold and Kidney Disorders	Seed husk
10	Kaith	*Acacia catechu*	Digestive. Appetizer, Astringent, and Antiseptic	Bark and heartwood
11	Kutki	*Picrorhiza kurrooa*	Antipyretic properties and against stomach disorders	Roots
12	Long pepper	*Piper longum*	Effective against Cough, Stimulant and Carminative	Flower
13	Quinine	*Cinchona ledgeriana*	Antipyretic, Astringent, Antimalarial	Bark
14	Tejpat	*Cinnamomum tamla*	Stimulant and carminative	Leaves

Source: Database on Medicinal Plants by CUTS Centre for International Trade, Economics & Environment, 2004

Table 4 : Active Ingredients In Common Medicinal Plants

S.N	Medicinal plant	Active ingredient/s	Method of propagation
1	Ashwagandha	Alkaloids and Withanolids	Seeds
2	Asparagus	Asparagose	Spears
3	Bishop's weed	Psoralens	Division of stem
4	Chiraita	Chirettine	Seedlings raised in nurseries
5	Chlorophytum	Saponins	Tubers
6	Clove	Eugenol	Seeds (it is a tree)
7	Coriander	Phenolic acid compounds	Seeds
8	Dandelion	Taraxacina (pectin)	Seeds
9	Fenugreek	Trigonellinelline	Seeds
10	Garlic	Allicin	Cloves
11	Ginger	Gingerols and Shogaols (terpenes and oleoresins)	Rhizomes
12	Isabgol	Quinoline	Seeds
13	Kutki	Kutkosides (Glucosides)	Seeds
14	Long pepper	Piperine	Seeds, suckers, cuttings
15	Mint	Menthol	Runners

Source: WHO Monographs on selected medicinal plants VOLUME 3, 2007

Bibliography

WHO, 2007, Monographs on selected medicinal plants VOLUME 3

Duke, James A (2008), Handbook of Medicinal Plants of the Bible, CRC Press Taylor & Francis Group

Panda H (2002), Medicinal Plant Cultivation and Their Uses, Asia Pacific Business Inc, Delhi

CUTS Centre for International Trade, Economics & Environment, 2004, Database on Medicinal Plants

www.ingramcontent.com/pod-product-compliance
Lightning Source LLC
Chambersburg PA
CBHW051824170526
45167CB00005B/2141